CAREERS IN
PHARMACEUTICAL RESEARCH

SOME OF TODAY'S MOST PROMISING CAREERS can be found in the pharmaceutical industry. Pharmaceutical researchers (also known as pharmaceutical scientists or pharmaceutical chemists) conduct research aimed at improving human health by developing new medications. They spend many hours in laboratories and with computers conducting experiments, analyzing the results, and preparing new drugs for manufacture. These professionals may find novel ways to extract substances from plants; synthesize new medications from chemicals; or design new therapies by combining existing medications in innovative ways.

Most research positions are found at large pharmaceutical companies, biotechnology companies, and multinational chemical concerns that manufacture medicines. There are also opportunities to work with small private research labs, government regulators, nonprofit medical charities, and firms that conduct clinical trials. Some researchers teach at colleges and universities while pursuing their own projects.

Some researchers specialize in the discovery of new medicines or on improving existing compounds. Scientists may also be engaged in testing new medications and gaining government approvals. Others oversee safety and logistical aspects for industrial-scale manufacturing of both prescription and over-the-counter drugs.

The number of pharmaceutical research professionals is expected to grow in the coming decade. While competition is tough for the top jobs, salaries are high for these professionals. Government statistics show the average annual salary for medical scientists (a group that includes pharmaceutical researchers) is almost $95,000. A survey by the American Association of Pharmaceutical Scientists (AAPS) found its members earn an average annual base salary of almost $150,000. When bonuses and other supplemental income are added, the typical AAPS member makes more than $180,000.

Would a pharmaceutical career be right for you? Technical training is needed – even entry-level positions require a four-year degree in the life sciences, but many positions require a doctorate. Personal trails are also important. Are you good at science and math? Can you analyze a situation logically to understand data, identify options, and determine the best solution? Do you have good organizational skills? Do you communicate well speaking and in writing? Do you want to make a significant contribution to the health and welfare of others? These traits will position you well to succeed in the research field.

If you have good analytical and interpersonal skills, you can enjoy a financially rewarding career in pharmaceutical research. A combination of training, hard work, affinity for science and medicine, and positive personal traits can help you achieve the personal and professional satisfaction that accompanies the role of a successful pharmaceutical scientist.

WHAT YOU CAN DO NOW

IF A CAREER AS A PHARMACEUTICAL RESEARCHER interests you, there are many steps you can take to explore the field while you are still in high school. A well-rounded, rigorous high school curriculum will help you get ready. Take advanced science and math classes including biology, calculus, chemistry, and physics. Good technology skills are highly useful, as medical scientists rely on technology to conduct research, analyze data, and report their results. Enhance your writing skills with English classes, as good communications skills will be needed for writing reports, grant proposals, and presentations. Computer software classes may be available at your school, from local colleges, or through online tutorials.

To learn more about the profession, read industry and professional publications – many of which are available online. Visit the websites of professional associations, such as the American Association of Pharmaceutical Scientists (AAPS), and find out about local chapters. Networking with those who are already employed in your chosen field helps you learn about scholarships and volunteer opportunities in your community. Another useful resource is the Pharma Foundation, which provides grants to help students pursue research projects.

Set aside some time to talk directly with a pharmaceutical researcher or medical scientist in your community. Discussing the pros and cons of a potential career with someone who lives that career each day provides a realistic view of the pros and cons of the discipline. Encourage them to be frank and honest. While becoming a pharmaceutical researcher can bring you a great deal of personal and financial satisfaction, it is also a challenging career with its own special issues to consider, so make sure you understand the full picture.

HISTORY OF THE PROFESSION

THE ROOTS OF THE MODERN PHARMACEUTICAL INDUSTRY began in the early 1800s. For centuries before then, medicines were compounded by hand – one dosage at a time – by local apothecaries who prepared products for individual customers in their pharmacies and laboratories. In the 1800s, some of those apothecaries began manufacturing and distributing botanical drugs for wholesale suppliers. Several modern pharmaceutical companies began as local apothecaries in North America and Europe, including Merck, Schering, Hoffman-Lauraceae, Upjohn, Eli Lily, Borough-Welcome and Abbott Laboratories.

In the United States, the industry first emerged around Philadelphia, Pennsylvania, from 1818 through 1822. A handful of chemical companies built factories in the area and began manufacturing pharmaceuticals. The first such company was started by Robert Shoemaker, who produced glycerin on an industrial scale.

The convergence of apothecaries and large-scale drug suppliers through the 1890s laid the foundation for the current multinational pharmaceutical industry. Competing companies turned their attention from simply manufacturing existing drugs to developing new ones. German firms were the first to work with academic laboratories that were conducting research to develop new pharmaceuticals, dyes, and other materials. These partnerships developed synthetic chemicals that could combat toxins, as well as bacteria, microbes, and other disease-causing organisms. That success soon led companies to build their own research programs to develop proprietary medicines without academic participation.

These industrial research centers drove demand for a new category of medical sciences and created the profession now known as pharmaceutical research. Research chemists had previously only extracted medicines from plants, such as quinine and morphine. By the dawn of the 20th Century, pharmaceutical researchers were also able to extract and isolate animal hormones (such as adrenaline) for human use. Research into coal tars and dyes also brought other medical applications. Pharmaceutical scientists soon devised numerous vaccines, including those for diphtheria and tetanus.

Throughout the 19th Century, the pharmaceutical industry had grown without federal government oversight. In the United States, food and medicine safety was in the hands of state regulators, and laws and standards varied widely. However, a number of scandals involving tainted vaccines in Germany and France in the 1890s helped spur global regulatory action. In 1906, the US Congress established agencies which would later become the Food and Drug Administration (FDA). These forerunners of the FDA were originally charged with ensuring that interstate foods and medicines were properly labeled.

Over time, the FDA would become responsible for the safety of food production, medicines, cosmetics, animal feeds, and other items – including approving new drugs for public sale. However, during the early 1900s, US chemists and researchers actually introduced relatively few new drugs, as there were not many large companies in this country engaged in complex manufacturing techniques. Most of the new drugs were then being discovered and commercialized by the large German chemical companies who monopolized innovation in the industry. US patent laws had always protected new medications, preventing competitors from copying the discoverer's original formula for several years after it was introduced. However, during this period, there were few new drugs discovered or created in the United States, so little patent activity occurred.

The two World Wars drastically changed the competitive landscape for the global pharmaceutical industry. A military blockade in the 1910s forced US scientists to learn and duplicate German processes for manufacturing such vital drugs as aspirin. That new production capacity helped launch hundreds of small companies that continued to produce medicines after the war. This innovative trend continued during World War II, as a US government emphasis on pharmaceutical research led to numerous breakthroughs – particularly the development of penicillin by 11 pharmaceutical companies working through the War Production Board.

The postwar period was a time of expanded pharmaceutical research, development, and growth around the world. Between 1940 and 1950, the American pharmaceutical industry consolidated from hundreds of small firms into a small group of large corporations. These companies were better able to focus their resources on pharmaceutical research and more in-depth development of new medications. This trend towards larger pharmaceutical companies would continue well into the new millennium, as a handful of multinational conglomerates continue to dominate the sector.

The last half of the 20th Century also saw increasing government regulation of the pharmaceutical sector. While these regulations have little direct effect on R&D (research and development of new products), they did influence the approval process for bringing new medications to market, clinical trials, and how medications are distributed. For previous generations, consumers could ask their pharmacist for virtually any medication they desired (except for certain controlled narcotics). That ended in 1952, when the US Congress passed the Durham-Humphrey Amendment dividing medications into prescription and over-the-counter drugs. Patients would now need to obtain a written prescription signed by a doctor to obtain certain drugs. Ten years later, another amendment gave the FDA expanded power to set regulations for clinical testing and approval for new drugs.

Over the next several decades, regulators would continue to introduce stricter testing and purity requirements for medications in the wake of several public health issues. At the same time, patients and their families often pushed the FDA to bring new life-saving medications to market more quickly, particularly those that had already won approval in Europe and Asia. The FDA continues to be challenged to balance public safety against helping ailing consumers. In 1984, the Hatch-Waxman Act opened the door for today's generic pharmaceutical manufacturing industry. Generic drugs are ones where the original patents have expired, enabling any company to legally manufacture the same drug without following clinical testing protocols.

Beginning in the 1980s and accelerating in the 2000s, the pharmaceutical industry has brought computer simulations into its research and testing procedures. Some new medications are "discovered" – extracted from plants, or combined with other materials – while others are synthesized (designed from other materials to create a new medication). In either case, computers have become valuable tools throughout the research process, particularly in mocking up potential outcomes of new medications and analyzing the results of clinical tests.

Today, the pharmaceutical industry continues to grow and prosper by creating an ever-expanding array of medications to treat most human ailments.

WHERE YOU WILL WORK

MOST PHARMACEUTICAL RESEARCHERS work for pharmaceutical firms, biotechnology corporations, private laboratories, government agencies, and universities and colleges. They may also be found at chemical manufacturing firms that produce drugs. Most researchers in the federal sector work for such agencies as the National Institutes of Health, the Food and Drug Administration, or US Pharmacopeia, while those in state government are found in similar regulatory units. Those working for universities typically teach classes as well as pursue their own research projects.

The American Association of Pharmaceutical Scientists (AAPS) reports that while competition is "fierce" for jobs in life sciences, most PhD scientists "will ultimately find jobs in academia, government, or the pharmaceutical or biotechnologies industries. While these jobs share many things in common, there are differences in compensation and benefits, job security, workplace culture, and career advancement."

About six percent of medical scientists (the federal category that includes most pharmaceutical researchers) work in the pharmaceutical and medicine manufacturing sector. Some 34 percent conduct research and development in the physical, engineering, and life sciences disciplines. Another 27 percent work for colleges, universities and schools; 15 percent for hospitals; and four percent at medical and diagnostic laboratories.

Over 100,000 medical scientists are employed, more than 40,000 working for scientific research and development services firms. Colleges, universities, and professional schools are the second largest employer at about 20,000, followed by general medical and surgical hospitals (15,000), and pharmaceutical and medical manufacturing (6,000).

The top paying industries are professional, scientific, and technical services providers; computer systems design and related services; the federal executive branch of government; and pharmaceutical and medicine manufacturing.

The AAPS reports these trends in its survey of its members. "The large majority of US full-time employees are employed in private industry, though this varies somewhat with education level," AAPS states "PhDs and PharmDs are more likely than average to be found in academia." The association states that almost 70 percent of its members outside of an academic setting work for commercial pharmaceutical manufacturers, with the others employed by government agencies or non-drug manufacturing companies. Over half of those in private industry are employed by large pharma companies.

California is the state with the highest employment of medical scientists at almost 23,000. Massachusetts is second with almost 10,000, followed by New York, Pennsylvania, and Maryland. The top paying states are Connecticut, California, New Jersey, Kansas, and Maine. Meanwhile, the metropolitan areas with the highest level of medical scientist employment are Boston, followed by New York City, Los Angeles, San Francisco, Seattle, San Diego, and Baltimore.

Pharmaceutical researchers and other medical scientists typically work full time in clean, modern offices and laboratories where they use state-of-the-art equipment and technologies to conduct experiments and analyze data. Medical scientists sometimes work with dangerous biological samples and chemicals, but they take precautions that ensure a safe work environment.

THE WORK YOU WILL DO

OVER THE PAST CENTURY, PHARMACEUTICAL RESEARCHERS have played key roles in discovering and developing new medications. Their work has helped save millions of lives and improved the quality of life for many millions more. Working as a pharmaceutical researcher provides countless opportunities to contribute to healthier lives for people around the world.

Pharmaceutical researchers are also known as pharmaceutical scientists, medical scientists, or analytical chemists, depending on their exact roles and employers. They work in laboratories researching potential new medications and finding new uses for existing medications. The role is sometimes confused with the similar-named pharmacist. However, pharmacists dispense existing drugs to patients, filling prescriptions written by doctors or guiding customers to over-the-counter medicines. Pharmaceutical scientists work in labs and offices to discover, design, develop, and test new treatments that – once approved by the FDA – can be manufactured and sold to the general public.

Most researchers work for large pharmaceutical and biotechnology corporations to develop and manufacture new drugs. Others are found in government regulatory agencies, chemical companies that manufacture drugs, nonprofit organizations that focus on specific diseases, and smaller biotech startups. Some may work at companies that simply run clinical trials or provide similar services for other corporations or government regulators. Pharmaceutical scientists may also become professors at colleges and universities, where they both teach and conduct their own research.

In a non-academic setting, pharmaceutical researchers typically start out as research assistants. They spend much of their time doing the routine work of collecting experiment data, recording results, tabulating and analyzing data, and writing reports on different aspects of the new drugs being studied. Much of their time is spent in front of a computer screen or in the laboratory, collecting samples or studying cells under a microscope. They also spend a considerable amount of time in meetings, presenting their research results or reviewing what other scientists have uncovered.

After gaining some experience on the job, research assistants can move up the career ladder to become full-fledged researchers, team leaders, or managers. While an entry-level assistant may be able to enter the profession with just a four-year degree in life sciences, researchers and executives are typically required to have a doctorate or at least a master's degree to advance their careers. For this reason, many researchers finish their bachelor's degrees, start out as assistants, and complete graduate studies while working in the lab.

The pharmaceutical sciences draw a broad range of scientific disciplines into the process of creating new medications. Developing new pharmaceuticals requires a team of scientists with backgrounds in many specialties. Those areas of expertise include:

- Analysis and data handling
- Biochemistry (developing vaccines and medicines from living organisms)
- Biopharmaceuticals (drugs from biological sources)
- Biotechnology (using biological processes for industrial purposes)
- Clinical pharmacology
- Drug metabolism
- Formulation design and development
- Manufacturing science and engineering

- Pharmacokinetics (the mechanics of drug actions)
- Pharmacodynamics (the study of how drugs move through the body)
- Pharmaceutical quality
- Regulatory sciences
- Veterinary sciences (medications for pets and other animals)

In most settings, pharmaceutical researchers are divided into small groups where each member focuses on one specific problem (or one small piece of a larger puzzle). The team typically includes representatives of several of the disciplines listed above, as no one person can be an expert on all phases of pharmaceutical science. Teams work together in a collegial atmosphere, pooling their skills and abilities to work towards their common goals. At most companies and universities, the research team has access to modern, sophisticated medical and computer technology to support their efforts.

Researchers generally focus on one of the three main phases in the pharmaceutical process: drug discovery, drug development, and manufacturing. The specific duties of an individual researcher will vary depending on which part of the process is their focus. The early stages are more research oriented, while manufacturing applies industrial practices to mass-producing a finished product.

Drug Discovery

The discovery phase covers identifying and designing potential new drugs. This is the most research-intense period, as pharmaceutical scientists may look at thousands of compounds to narrow down the list of candidates to a few good prospects. Discovery can refer to designing and creating an entirely new drug, or to extracting substances from existing sources (such as plants). Discovery also covers finding new uses for current medications and creating new therapies – for example, combining doses of two or more existing drugs to see how they interact in hopes that the combination will produce better patient outcomes. Research institutes and universities have traditionally conducted much of the early-stage drug discovery, although a growing amount of that work is being done by startups and multinational pharmaceutical corporations with vast research and development (R&D) resources.

Drug Development

Once a compound is discovered or a new therapy is identified as a potential medication, the next step in the process is drug development – determining the drug's suitability for patients and working out the details of how it will be made. Development includes determining the right doses and formulas of a proposed new drug that will achieve the desired effects in patients. Safety is a vital component of the development phase.

Research during the development stage often includes *in vitro* studies (conducted on microorganisms or on cells removed from an organism), *in vivo* studies (experiments conducted on living creatures, typically animals), *in silico* studies (computer simulations), and ultimately, *clinical trials* involving human participants. Researchers normally work with physicians who will administer the new drug to people who agree to participate in clinical trials. They then track the health of participants, looking for signs of improved health as well as potential side effects from the new medication.

These formal studies plus a large volume of additional data are filed as part of the regulatory review process. Researchers create data summaries, graphs, charts, and reports that are submitted to the FDA as part of the New Drug Application (NDA) process. The development phase can require 20 years or more, and incur significant costs, so drug development is typically done by large pharmaceutical and biotech companies with vast financial resources.

Drug Manufacturing

Once the FDA approves the drug, the next step is manufacturing the new medication on an industrial scale for distribution across the country or around the world. Pharmaceutical and medical scientists may have a role in designing quality and safety protocols around the manufacturing process. However, chemical and materials scientists are typically more involved at this phase, along with industrial designers, marketers, and logistics experts.

The basic duties of pharmaceutical researchers are similar throughout these phases of bringing a new drug to market:

- Gather, analyze, and interpret complex medical data.
- Design and conduct studies that investigate human (or animal) diseases, as well as attempting to treat or prevent those diseases.
- Prepare and analyze medical samples or other data to better understand pathogens, chronic diseases, or toxicity.
- Design experiments to determine the effects of medications on the human body.
- Use computers to synthesize drug molecules, and simulate their interaction with other molecules or organisms.
- Evaluate the effectiveness and safety of new drugs.
- Communicate the results of research through written or spoken presentations.

- Create and test medical devices related to the pharmaceutical industry.
- Determine standards for drug potency, dosage levels, and methods to administer the drugs.
- Write research grant proposals and apply for government or private funds (typically in a university or small company setting).
- Document research results that may be published in medical journals or presented at industry conferences.

While the range of duties can be broad, the depth of knowledge required is extensive. Most pharmaceutical and medical scientists specialize in one particular area of research where they can gain in-depth knowledge of current medical capabilities and identify the areas of greater need for new research. Those specialized roles may include:

- **Cancer researchers**
- **Clinical and medical informaticians,** who focus on statistical analysis of large databases.
- **Clinical pharmacologists**, concentrating on the clinical aspects of drug development.
- **Gerontologists,** who address the needs of an aging population.
- **Immunologists,** concentrating on how chemicals and drugs affect the immune system.
- **Research histologists,** who focus on aspects of human tissue.
- **Serologists,** who research bodily fluids such as blood.
- **Toxicologists,** who are concerned with poisons, hazardous chemicals, and other harmful substances.

While most pharmaceutical researchers work for private industry, others work in academic or government settings. Many doctoral students pursue a position with a college or university that allows them to teach as well as conduct research into areas they find most interesting. Those who become tenured faculty enjoy job security and professional standing. They can also work as consultants, serve on advisory boards, and start companies to commercialize their own research.

Meanwhile, the number of government positions continues to grow, including both regulatory branches of government and agencies conducting pure research. Agencies such as the FDA, National Institutes of Health, the Centers for Disease Control, and Department of Defense employ life scientists from various disciplines. While recent cutbacks in government spending have lessened the number of permanent positions, many agencies increasingly use contract employees to conduct their pharmaceutical and medical research.

A background in pharmaceutical research can also prepare you for a variety of non-research career options. Many veteran scientists have moved into related fields where they can leverage their knowledge of the pharmaceutical research process to make other contributions to the industry. Those areas include regulatory compliance, medical affairs, pharmaceutical marketing, and clinical research management. Even if you do not spend your entire career in a white lab coat and latex gloves, your work as a pharmaceutical researcher can open doors to a wide array of employment choices.

STORIES OF WORKING PROFESSIONALS

I Am an Analytical Chemist in Pharmaceutical Research

"I became interested in the medical field at an early age, but felt I really wasn't drawn to treating patients as a doctor or working in a hospital. I first learned about the pharmaceutical industry at a high school career fair, and decided that research to help find new medications was a better career path for me.

In my university studies in upstate New York, I concentrated on chemistry. I received my bachelor's degree in chemistry, followed by both a master's and my PhD in analytical chemistry. I spent the next 11 years working in analytical R&D for a variety of pharmaceutical and biotechnology companies. I progressed from research assistant to senior scientist while working on all phases from discovery through clinical trials. Those years included helping develop medications to address lipid storage diseases, colitis, and other conditions.

I came to my current position as an analytical chemist four years ago at a biotechnology and pharmaceutical corporation near Boston. Our company focuses on diagnosis and treatment of genetic ailments related to carbohydrate metabolism. I provide analytical support to the discovery phase for possible new therapeutics. During drug development, I provide analytical support for clinical trials until approval of the new drug application by the FDA.

Part of my day is spent managing the work of another scientist who performs various characterization, ingredient stability, and impurity tests on new substances. I often perform additional chemical testing of my own. I am also involved in helping optimize our methodologies and validating those changes to procedures. My role includes working with our manufacturing facilities to make sure all the appropriate methods for creating new drugs are communicated and implemented. I help make sure that the information we communicate to the FDA and other regulators is complete and accurate.

My favorite parts of the job involve keeping up-to-date with new scientific trends and techniques, and devising more efficient strategies to tackle the many issues we commonly encounter in pharmaceutical development. While I enjoy my current work, it is reassuring that an influx of new biotech companies in the Boston area means there are always plenty of opportunities for new challenges."

I Am a Pharmaceutical Researcher

"I prepared for my career as a research scientist in the pharmaceutical industry by obtaining my bachelor's and doctoral degrees in biology at the University of Birmingham, England. I also worked as a research assistant at the university while completing my PhD, providing additional on-the-job training and experience.

After graduation, I moved to the United States and gained employment with a large pharmaceutical company in California. My work is in the discovery phase. Our project team seeks new antibiotics to better treat serious infections that can lead to hospitalization. Our team includes experts from several disciplines, including biochemistry, pharmacology, drug metabolism, and microbiology. We pool our skills and knowledge to identify possible candidates for new drugs. The odds are long – maybe one in 10,000 compounds ultimately becomes an approved new medication – and it can take 10 years to bring a new product to market. But when we succeed, it makes all our hard work meaningful.

I also lead a small team that conducts experiments on compounds created by the chemists in the larger project group.   We follow a variety of laboratory procedures that determine the microbiological properties of these new compounds. Our primary focus is determining whether the potential new drugs will actually inhibit bacterial growth. Our experiments, along with other research and testing procedures in other disciplines, determine whether the compound has the properties we seek. If so, we will pass those compounds on to another set of scientists who are responsible for the development process.

The best part of my job is knowing our team is making a vital contribution to our company's efforts to find new drugs to cure people with serious illnesses. A couple of times each year, our company reminds us of the ultimate goal by bringing in patients whose lives are better because of our company's medications. It is truly inspiring to see the difference my work makes in the lives of those people."

PERSONAL QUALIFICATIONS

A SUCCESSFUL CAREER AS A PHARMACEUTICAL RESEARCHER requires technical knowledge as well as positive personal characteristics. Basic math, science, computer, and analytical skills provide the foundation for understanding and evaluating complex laboratory results. You can acquire these technical skills in high school and college, and then enhance that knowledge with on-the-job experience and training. Often, you will be working on your own, but you will also likely be a member of a team, so you must be capable of working with a group of colleagues when necessary.

Several personal qualities are important to help you perform well on the job – as well as endure the ups and downs of a research career. Honesty and integrity will be critical, as your work potentially affects the health and welfare of thousands of people. Sound judgment, critical thinking, problem-solving techniques, and decision-making skills also come into play as you analyze results and scrutinize data to find the right answers. Being able to keep things organized and managing your time efficiently will improve your productivity.

You also need to be detail oriented to unlock the answers buried within huge quantities of information. Your work will include observing results, analyzing data, spotting trends, tweaking experiments, and repeating the process until you are satisfied with the outcome. Much of that analysis and documentation occurs digitally, so computer skills will help you succeed. Staying up to date on current trends in the industry also help you learn from the work others are doing in your area of specialization.

Following all the rules in a research lab is critical. Government regulations affect many aspects of the pharmaceutical industry, so you must understand and follow those rules. You will also need to learn and comply with the rules and procedures at your company for product safety, laboratory controls, documenting results of experiments, and similar activities. Lab safety is particularly important when there are hazardous materials or infectious disease samples in the workplace.

Solid communications skills are vital. You must be able to communicate well in writing, as you will often submit lengthy reports, write emails, and share progress with colleagues. Speaking effectively is important for talking with peers, presenting your findings, and meeting with executives. Whether you are talking with your lab partner or the vice president of research, you must be able to make your points understandably, and be an active listener who can detect the concerns of others.

The pharmaceutical industry can be demanding, with long hours spent in a laboratory and pressure to produce results to improve people's lives. You will need to be able to remain calm as experiments unfold over time, and persevere through periods of trial and error. It can be challenging to stick with months or years of research to bring a single new medication to market. If you have the desire to make a contribution to the health and well-being of others by developing breakthrough therapies and medicines, the rewards can be immeasurable.

ATTRACTIVE FEATURES

PHARMACEUTICAL RESEARCH AND medical science is a stable field offering a wide variety of opportunities to those entering the profession. The earnings, professional rewards, and personal potentials for this career path continue to grow as the pharmaceutical industry develops new medicines and therapies to address a variety of disorders. A vast range of options exist for building a career at large public multinational corporations, private research laboratories, government agencies, educational and research centers, and nonprofit institutions.

If you like solving problems, thinking creatively, enjoy working with technology, interacting with peers, and generally making life easier for others, you should find this career fulfilling. You will work with (and learn from) the top experts in medical research as you continually improve the quality of your company's products and services. Your achievements will have a positive impact on your organization, your customers, your community – and people around the world who benefit from new medicinal breakthroughs. Success can bring personal fulfillment, professional growth, and the satisfaction that comes from a job well done.

Companies that employ pharmaceutical researchers provide competitive salaries, attractive benefits, state-of-the-art technology, technical training, and numerous perks to help attract the best candidates to their workplace. Paid vacations and holidays, generous retirement packages, performance bonuses, and the ability to acquire company stock at a discount are frequent benefits for corporate researchers. These organizations also typically offer employees a well-defined path towards career advancement, either as a more senior researcher or in a management position.

Earnings are excellent, often topping $150,000 annually and peaking close to $200,000.

Pharmaceutical researchers are held in high esteem by peers and colleagues throughout the industry, as well as patients who reap the benefits of their efforts. Researchers work closely with other scientists to pursue important work that benefits people in their time of need. You will be working in modern, clean offices and labs using the best technology on a daily basis.

You may decide to climb the corporate ladder, join a university faculty or research facility, or use your skills to focus on a specific illness by working with a nonprofit organization. Whichever path you choose, a career in pharmaceutical research can be both financially and personally rewarding for many years to come.

UNATTRACTIVE ASPECTS

WHILE A CAREER IN PHARMACEUTICAL RESEARCH can be rewarding, it is also likely to be demanding and stressful. The pharmaceutical development life cycle is a long, repetitive process where the rate of progress can seem slow as you continually repeat and tweak experiments. You will spend much of your time designing experiments, observing results, analyzing data, proposing new approaches, and revisiting previous work. Failure is frequent – only a small percentage of potential drugs actually make it outside the laboratory.

Pharmaceutical research is complex and sometimes exhausting. Researchers often work long hours, nights and weekends, particularly around project deadlines. Scientists are paid a set annual salary rather than an hourly wage, so there is no overtime pay. Projects to develop new medications often last for many years. Spending extended hours in a laboratory for all that time can become frustrating. Long periods of overwork may cause personal stress and sometimes burnout. As with any career, you may have to deal with demanding managers, personality conflicts, government regulators, and office politics.

Pharmaceutical professionals also face constant demands for education – both to get started in the field and to keep up with new industry developments. Some employers offer training courses at the entry level, or when new techniques and processes are introduced into the workplace. However, for other continuing education needs, many employers expect you to schedule (and possibly pay for) your own training. Even when the company offers classes during working hours, you are responsible for proceeding with your normal job duties. You may have to work extra hours to ensure your regular work does not fall behind.

Competition for the top research jobs continues to be tough. There are a limited number of corporate jobs and a large number of qualified applicants. With more American medical manufacturing heading overseas, there are fewer opportunities to find employment in some sectors as jobs are outsourced. At the same time, the number of academic research and teaching positions continues to be limited, particularly for tenure-track professors.

In general, researchers work in clean, safe, modern office environments. However, there can be risk of injury if hazardous or infectious materials are not stored and handled properly. Scientists may be exposed to health or safety hazards when handling certain chemicals. However, proper procedures, such as wearing protective clothing when handling hazardous chemicals will usually protect you.

Overall, long hours, stress, and difficult work conditions can be taxing on pharmaceutical researchers. In particular, frequent computer use can lead to eyestrain, back pain, and repetitive motion injuries such as carpal tunnel syndrome.

EDUCATION AND TRAINING

GETTING STARTED AS A PHARMACEUTICAL SCIENTIST requires at least a four-year college degree in the life sciences, medical, or chemistry fields to land an entry position. However, becoming a researcher, executive, or professor will typically require a doctorate, master's degree, or similar graduate education.

Almost one third of all medical scientists (a group that includes pharmaceutical researchers) have a doctoral degree. A survey by the

American Association of Pharmaceutical Scientists (AAPS) found that half of their members hold a PhD, five percent have the Doctor of Pharmacy (pharma), 19 percent a master's, and 14 percent a bachelor's degree. Members reported concentrations in chemistry, pharmacokinetics, pharmaceutical/medicinal chemistry, analytical chemistry, and biochemistry.

Typical basic courses in these undergraduate curricula cover anatomy, biology, biochemistry, math, writing, and physics. More advanced classes delve into details about pharmacology and pharmaceuticals. Other courses may include immunology, physiology, toxicology, microbiology, or veterinary medicine.

After completing their four-year bachelor's degree, most future pharmaceutical researchers and other medical scientists enter a PhD program. Many choose dual-degree programs that include the PhD plus a specialized medical degree, such as Pharma. The PhD studies focus on research methods, whereas students in dual-degree programs learn both the clinical skills needed for clinical medicine and the research skills needed to be a laboratory scientist.

The first two years of a PhD program generally include pharmacology, psychology, pathology, medical ethics, and medical law. Medical scientists often continue their education with post-doctoral work, where the extra independent lab work provides experience in specific processes and techniques, such as gene splicing, which is transferable to other research projects.

Which college should you attend for your undergraduate and graduate work? There are several national surveys that rank the best schools for chemical and biological scientists. For example, US News and World Report lists the top schools in the country for an undergraduate degree:

- Massachusetts Institute of Technology
- University of California-Berkley
- Stanford University
- University of Texas-Austin
- University of Minnesota-Twin Cities
- Georgia Institute of Technology
- University of Wisconsin-Madison
- California Institute of Technology
- Princeton University
- University of Delaware

The magazine's ranking of the top graduate schools for chemistry (including those that prepare students to work for pharmaceutical companies) includes many of those same schools. Cal Tech tops the list, followed by MIT, UC-Berkeley, Harvard University, Stanford, Illinois-Urbana, Northwestern, Scripps Research Institute, Wisconsin, Columbia, and Cornell.

There is also this ranking of the top global universities for biology and biochemistry. These US schools made the list:

- Harvard
- MIT
- UC-Berkeley
- Stanford
- University of California-San Francisco
- UC-San Diego
- University of Oxford
- University of Cambridge
- John Hopkins University
- UCLA

The top graduate schools for biological sciences also include many of those schools: Harvard, MIT, Stanford, UC-Berkeley, Cal Tech, John Hopkins, UC-San Francisco, Yale University, Princeton, Cornell, Duke University, and Washington University in St. Louis.

Once you land that new job, expect your training to continue. Most companies that hire researchers offer in-house programs to enhance your basic knowledge, teach you about their company, and share up-to-date knowledge about the markets and industries they serve.

Some employers may require candidates for clinical research associate jobs to complete an accredited clinical research assistant program. These programs cover such topics as researcher roles and responsibilities, and key concepts of the industry.

Pharmaceutical researchers and other medical scientists typically do not need licenses or certifications. Those who administer drugs, gene therapy, or otherwise practice medicine on patients in clinical trials need a license to practice as a physician. Similarly, anyone who distributes pharmaceuticals to the general public must be licensed as a pharmacist. Continuing education is required to renew a pharmacist license through accredited organizations. AAPS also provides continuing education classes for researchers on such topics as regulatory affairs and new technologies.

In addition to formal training, you can also expect to spend a great deal of time simply staying up to date with current news and industry developments. You will regularly review research reports and similar documents that focus on your sector. Government regulations, market trends, and product safety concerns change constantly, so plan on setting aside a significant amount of time simply keeping pace with the rapidly-changing pharmaceutical industry.

EARNINGS

THE MEDIAN ANNUAL EARNINGS among medical scientists (a group that includes most pharmaceutical researchers) are about $85,000. The top 10 percent of medical scientists earn over $150,000 annually, while the bottom 10 percent receive about $45,000.

Scientists working in pharmaceutical and medicine manufacturing earn about $110,000.

Some pharmaceutical researchers are classified as chemicals and materials scientists, depending on their exact duties with their employers. The median pay for chemists is a little more than $70,000, while materials scientists earn over $90,000. Chemists working in pharmaceutical and medicine manufacturing earn on average $70,000.

Even higher salaries are reported in a recent survey of members of the American Association of Pharmaceutical Scientists (AAPS). Almost 20 percent of its nearly 10,000 members report an average annual base salary of almost $150,000 in the United States, an increase of almost three percent over the previous survey. When supplemental income (such as bonuses and consulting work for other employers) is added to the base pay, members earn an average of over $180,000.

Another study by Monster.com found that the position of biotechnology senior research scientist is among the top-paying jobs in the pharmaceutical industry. These researchers earn a media annual salary of almost $100,000. Individuals with five to 10 years of experience begin bringing in $97,000, whereas those early on in their careers start at $90,000, according to the website. Another top position is research scientist, with median earnings of $82,000.

OPPORTUNITIES

EMPLOYMENT FOR PHARMACEUTICAL RESEARCHERS and other medical scientists is projected to continue growing through the foreseeable future. Employment is projected to climb eight percent within the coming decade. The sector is expected to add 9,000 jobs to the almost 110,000 positions currently reported.

An aging population, increased rates of chronic health conditions, and a growing reliance on pharmaceuticals are all factors in the increase. New discoveries and techniques should open frontiers in research to further drive demand for scientists who can help treat diseases and contain international epidemics.

The federal government continues to provide the major source of funding for medical research.

Meanwhile, jobs for chemical and material scientists (another group that includes pharmaceutical researchers) are expected to increase somewhat more slowly as manufacturing industries overall continue to decline. Some pharmaceutical companies may continue to increase the outsourcing of their R&D activities, rather than doing the research in-house. This outsourcing strategy is likely to cause faster growth in the employment of chemists in small, independent research-and-development firms, rather than in the more traditional large manufacturers. However, large biotechnology and pharmaceutical firms, along with colleges and universities, are still expected to provide employment for research laboratory workers, particularly those with a PhD.

Pharmaceutical research positions can be found in cities of all sizes across the country, although most job openings tend to be concentrated in large metropolitan areas. The state with the highest level of employment of medical scientists is California with about 23,000 jobs. Massachusetts is second, followed by New York, Pennsylvania, and Maryland. The metropolitan areas with the highest levels of employment are located in Texas, California, New York, Vermont, Massachusetts, and Connecticut.

California is also the top employing state for chemists (including medical chemists, who research and develop chemical compounds that can be used as pharmaceutical drugs). Texas ranks second, followed by Pennsylvania, New Jersey, and New York. California also leads employment of materials scientists, followed by Ohio, Massachusetts, New York, and Pennsylvania.

Pursuing a career in pharmaceutical research does not necessarily mean moving to a large city or a new state. Many smaller firms have offices in mid-sized cities. Colleges and universities are found in communities of all sizes. Also, a growing number of companies allow employees to telecommute, working from home rather than making a daily trip to company headquarters. Companies generally benefit from higher employee satisfaction, as well as reducing their operating costs, when employees work from home.

GETTING STARTED

ARE YOU READY TO PURSUE A CAREER as a pharmaceutical researcher? The first step is learning more about the profession. Begin by gathering more detailed information about the pharmaceutical industry and the research discipline. Determine which sector interests you most. Would you rather work with a large multinational pharmaceutical firm, or at a smaller research lab? Does government regulation and food safety interest you? Or would you prefer a not-for-profit institution seeking a cure for a deadly disease? You may be able to find these opportunities in your community, or you may decide to move to another area that offers a greater diversity of job options.

Next, determine how you can begin working towards your goal. Pursuing a four-year degree in the life sciences is a required first step to land an entry-level position as a research assistant. After graduation, do you want to continue immediately with a graduate program? Or will you first spend a few years working as a researcher while pursuing your graduate degree? Your choice of employer will also help determine what level of education you will need to get started and how to best advance in your career.

Once you identify the training you will need, look for colleges and universities that provide the appropriate education. Give some thought to how you might gain entry-level experience while still in school that would help you land your first job. Are there internships in medical science and research available (preferably in pharmaceuticals)? You may be able to find a relevant part-time position to help enhance your résumé while earning a salary. Any job – even working in a drug store pharmacy – can provide valuable experience while demonstrating your commitment to the industry.

Research the profession. Written information about the pharmaceutical sector can be found at libraries, through professional and trade associations, from colleges and universities, and through the guidance counselors at your high school. Other sources include the Internet, industry trade magazines, and the business sections of national magazines.

Spend some time talking to people who already work in the profession. Ask them what personal qualities, skills, and training are most beneficial when pursuing that first paying position. You can find these individuals through professional associations, family networks, and by contacting drug companies or hospitals. Industry associations may also provide contacts, along with information on internships, training programs, and networking venues.

Discuss your plans with family and friends to get input on whether this sounds like an appropriate career choice for you. Your school counselor can share helpful information about local opportunities.

Once you gather your data, it is time to give careful thought to whether pharmaceutical research still seems like a good choice for you. Do you enjoy science – particularly chemistry and biology classes? Are you ready to spend many years in college and graduate school logging countless hours of lab work in the life sciences? Are you comfortable dealing with numbers, statistics and mathematical spreadsheets? Can you handle working long hours and the stress of pressure to perform experiments quickly and efficiently? Do you communicate well in writing and speaking? Would you like to help improve the health and well-being of potentially thousands of people?

The most important determination is whether you can see yourself happily pursuing a successful career in the pharmaceutical sciences. If so, start taking your first steps today towards a rewarding, fulfilling career!

ASSOCIATIONS

■ **Academy of Pharmaceutical Sciences**
www.apsgb.org

■ **American Association of Pharmaceutical Scientists**
www.aaps.org

■ **American Society for Clinical Pharmacology and Therapeutics**
www.ascpt.org

■ **American Society for Pharmacology and Experimental Therapeutics**
www.aspet.org

- Canadian Society for Pharmaceutical Sciences
www.cspscanada.org

- New Jersey Pharmaceutical Association
www.njphast.org

- Parenteral Drug Association
www.pda.org

- Pharmaceutical Management Science Association
www.pmsa.net

- PhRMA Foundation
www.phrmafoundation.org

PERIODICALS

- AAPS Newsmagazine
www.aaps.org

- British Journal of Pharmaceutical Research
www.sciencedomain.org/journal/14

- European Journal of Pharmaceutical Sciences
www.sciencedirect.com

- Journal of Applied Pharmaceutical Science
www.japsonline.com

- Journal of Pharmaceutical Sciences
www.jpharmsci.org

- Journal of Pharmacy Practice and Research
www.onlinelibrary.wiley.com

- Pharmaceutical Research
www.aaps.org/PharmRes

- Pharmacognosy Research
www.phcogres.com

- The Pharmacologist
www.aspet.org

- World Journal of Pharmaceutical Sciences
www.wjpsonline.org

WEBSITES

■ **AAPS Career Center**
www.aaps.org/careers

■ **Accreditation Council for Pharmacy Education**
www.acpe.org

■ **Pharmaceutical Research and Manufacturers of America**
www.pharma.org

■ **Pharmaceutical Scientist Careers**
www.explorehealthcareers.org

■ **PhRMA Foundation**
www.phrmafoundation.org

Copyright 2017 Institute For Career Research
Careers Internet Database Website www.careers-internet.org
Careers Reports on Amazon
www.amazon.com/Institute-For-Career-Research/e/B007DO4Y9E
For information please email service@careers-internet.org

www.ingramcontent.com/pod-product-compliance
Lightning Source LLC
Chambersburg PA
CBHW070720210526
45170CB00021B/1391